Maths Sticker Workbooks
Division

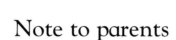

Wendy Clemson and David Clemson

Note to parents

This book is part of a programme of workbook titles that are designed to support school-work and make learning fun. Each page demonstrates a different mathematical concept and provides example calculations. When your child has met each concept and understood it, the practice calculations can be tried.

Many of the questions and puzzles in **Division** are answered with stickers, which are found on the middle two pages of the book. Your child will need plenty of scrap paper to write down the calculations featured on each page before he or she can start to work out the answers. All answers are provided on page 16.

How to use this book:

 The star sign means there is a sticker to put here on the page.

 Wherever your child needs to fill in an answer, there is a blue box like this one to write in.

 The calculator picture appears whenever a calculator is needed to solve a puzzle.

 In the top left-hand corner of each page there is a space for a "reward" sticker. Your child can add it when he or she has completed the puzzles.

DORLING KINDERSLEY

LONDON • NEW YORK • MUNICH • MELBOURNE • DELHI

Dividing numbers

Division is repeated subtraction, taking away again and again. Study the number line to see how division and subtraction connect and then practise dividing numbers with these fishy calculations.

Number line
Use this number line to see how division and subtraction are linked.

To work out how many twos are in 10, you could do the following:

10 - 2 - 2 - 2 - 2 - 2 = 0

There are 5 twos in 10.

Signs and symbols
Division calculations are usually written out using signs and symbols, as shown below. Use these signs and symbols to set out your division calculations.

10 ÷ 2 = 5

This sign means divide by. This sign means equals. This number is the answer.

Dividing groups of fish
A pet shop has 18 tropical fish. Solve these puzzles and find the correct sticker answers.

If the pet shop has 3 aquariums, how many fish fit equally into each one?

If half the fish are sold, how many fish will be left?

If the fish were shared out between 6 children, how many fish would each child receive?

If the fish were sold in pairs, how many pairs could you buy?

If 3 fish are sold, how many fish will be left in each of the 3 aquariums?

2

Leaping fish

Archerfish leap out of the water to catch spiders to eat. If there are 21 spiders and 7 fish in a pond, each fish can have 3 spiders. How many spiders can each fish have in these puzzles?

28 ÷ 7 = ☐
Spiders — Fish in the pond — Spiders per fish

12 ÷ 3 = ☐
Spiders — Fish in the pond — Spiders per fish

30 ÷ 10 = ☐
Spiders — Fish in the pond — Spiders per fish

48 ÷ 8 = ☐
Spiders — Fish in the pond — Spiders per fish

Shellfish supper

A group of 60 shellfish can be shared out amongst a varying number of lion fish. Study the example below and then find the correct sticker answers to solve the calculations.

15 shellfish each for **4** lion fish.

10 shellfish each for lion fish.

30 shellfish each for lion fish.

6 shellfish each for lion fish.

12 shellfish each for lion fish.

20 shellfish each for lion fish.

Division practice

Solve these division calculations and fill in the boxes with the correct answers.

30 ÷ 5 = ☐

35 ÷ 7 = ☐

4 = ☐ ÷ 9

16 = 32 ÷ ☐

12 = ☐ ÷ 2

4 = ☐ ÷ 4

Division practice

Use your division knowledge to see how division and multiplication are connected. Then try these sporty puzzles to test what you know about division.

Relay teams

Multiplication and division are closely connected. We say multiplication is the inverse of division. Study the relay team puzzle to see how they connect.

Can you see how the calculations connect?

There are 20 children to be put into teams of 4 for the relay. How many relay teams is that?

20 ÷ 4 = 5

To check your calculation is correct, try it the other way around.

5 x 4 = 20

Team players

Imagine you are a sports coach. Use the key below you to help you solve these sticker puzzles.

Sport	Number of people in each team
Basketball	5
Hockey	11
Volleyball	6

You can put 66 team players into how many hockey teams?

How many volleyball teams can you put 66 team players into?

You can put 55 team players into how many basketball teams?

How many hockey teams can you put 55 team players into?

Tennis balls

Each pack contains 6 tennis balls. Share out the packs and solve these puzzles.

12 ÷ 6 = ☐ packs

42 ÷ 6 = ☐ packs

24 ÷ 6 = ☐ packs

6 packs = ☐ ÷ 6

3 packs = ☐ ÷ 6

9 packs = ☐ ÷ 6

Match tickets
A family has 12 tickets for a football match.

If the family shares out the tickets with other families, how many will each family have?

12 ÷ 6 = ☐
Tickets   Families   Tickets per family

12 ÷ 3 = ☐
Tickets   Families   Tickets per family

12 ÷ 2 = ☐
Tickets   Families   Tickets per family

If they share out the tickets as shown below, how many families will have a share?

1 = 12 ÷ ☐
Ticket per family   Tickets   Families

3 = 12 ÷ ☐
Tickets per family   Tickets   Families

Sports equipment
Imagine you are a sports coach and have 24 pieces of sports equipment to be shared out between different numbers of people. Complete these divisions.

24 ÷ 2 = 12
24 ÷ 6 = ☐
24 ÷ 8 = ☐
24 ÷ ☐ = 8
24 ÷ ☐ = 6
24 ÷ ☐ = 4

Basketball scores
Below is a list of baskets scored by the star players of 3 teams. Find the stickers to show how many stars each team has.

Team one scores:
12 baskets • 6 by each star player   ☆ Stars

Team two scores:
18 baskets • 6 by each star player   ☆ Stars

Team three scores:
8 baskets • 2 by each star player   ☆ Stars

Remainders

Sometimes when we do a division there are numbers left over. These are called remainders. Do the factory puzzles to work out some remainders.

Stereo puzzles

A factory is supplied with all the parts to make a personal stereo. If extra parts are supplied, there will be a remainder. Follow the calculation to the right to see how a remainder is worked out.

4 stereo control buttons are needed for each stereo. If 5 are supplied the remainder is 1.

$$5 \div 4 = 1 \text{ remainder } 1$$

Supply | Buttons per stereo | Stereo | Button left over

Study the puzzles below and work out how many stereos can have a set of 4 buttons and how many will be left over. Fill in the boxes.

17 Buttons
Stereos ☐   Remainder ☐

23 Buttons
Stereos ☐   Remainder ☐

9 Buttons
Stereos ☐   Remainder ☐

14 Buttons
Stereos ☐   Remainder ☐

25 Buttons
Stereos ☐   Remainder ☐

35 Buttons
Stereos ☐   Remainder ☐

Cassette tapes

Each stereo is sold with 3 cassette tapes. How many tapes will be left over in the following calculations? Find the correct sticker answers.

16 tapes can be divided between stereos remainder

19 tapes can be divided between stereos remainder

Transporting stock

The stereos are packed in boxes, and a truck can carry 100 boxes. How many boxes will be left behind at the factory in the calculations below? Solve the puzzles and find the correct sticker answers.

202 boxes will fit into trucks  remainder

403 boxes will fit into trucks remainder

222 boxes will fit into trucks remainder

Factory calendar

The factory work is in weekly (Monday–Friday) shifts.

How many complete working weeks are there in the month below?

How many working days remain?

How many complete working weeks would there in a 31-day month?

How many working days would remain?

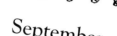

September

Sun	Mon	Tues	Wed	Thur	Fri	Sat
	1	2	3	4	5	6
7	8	9	10	11	12	13
14	15	16	17	18	19	20
21	22	23	24	25	26	27
28	29	30				

Working week

Each factory employee works 20 days per month. A working week is 5 days. How many extra days will these employees work in the following puzzles?

6 days a week for 4 weeks:

 full weeks and extra days.

5 ½ days a week for 4 weeks:

 full weeks and extra days.

5 days a week for 2 weeks and 5 ½ days a week for 2 weeks:

 full weeks and extra days.

7

Long division

Make fast work of dividing large numbers by using long division. Study the examples on this page and then try to solve the long division puzzles.

Dig divisions

An archaeologist on a dig has collected a box of 345 animal bone fragments. By using long division, she can divide the bones equally into several groups. Follow the calculations below to see how she can do this.

$345 \div 6 = ?$

Set out the calculation like a long division.

$$6 \overline{)345}$$

The answer will go here.
The working out will go here.

First, divide 6 into 34.

$$\begin{array}{r} 5 \\ 6\overline{)345} \\ 30 \\ \hline 4 \end{array}$$

6 goes into 34 five times.
Five times 6. Write 5 on the answer line.
The remainder is 4.

Then, divide 6 into 45.

$$\begin{array}{r} 57\,r3 \\ 6\overline{)345} \\ 30 \\ \hline 45 \\ 42 \\ \hline 3 \end{array}$$

Bring the 5 down next to the 4.
6 goes into 45 seven times remainder 3.
Write 7 remainder 3 on the answer line.

$345 \div 6 = 57\ r3$

$345 \div 11 = ?$

Now, follow this long division.

$$11 \overline{)345}$$

The answer will go here.
The working out will go here.

First, divide 11 into 34.

$$\begin{array}{r} 3 \\ 11\overline{)345} \\ 33 \\ \hline 1 \end{array}$$

11 goes into 34 three times.
Three times 11. Write 3 on the answer line.
The remainder is 1.

Then, divide 11 into 15.

$$\begin{array}{r} 31\,r4 \\ 11\overline{)345} \\ 33 \\ \hline 15 \\ 11 \\ \hline 4 \end{array}$$

Bring the 5 down next to the 1.
11 goes into 15 one time remainder 4.
Write 1 remainder 4 on the answer line.

$345 \div 11 = 31\ r4$

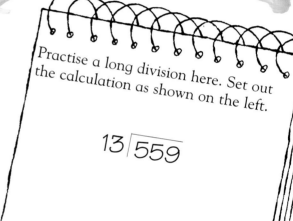

Practise a long division here. Set out the calculation as shown on the left.

$$13\overline{)559}$$

What is the answer?
$559 \div 13 =$ ☐

Now try to solve these long divisions.

$350 \div 25 =$ ☐

$442 \div 17 =$ ☐

2/3 Dividing numbers
Dividing groups of fish

5 6 9 3 9

Shellfish supper

2 3 6 5 10

4/5 Division practice
Team players

11 6 5 11

Basketball scores

4 2 3

6/7 Remainders
Cassette tapes

1 6 5 1

Transporting stock

2 22 2
3 2 4

Working week

4 2 4
4 4 1

9 Approximation
Missing numbers

Approximately 2
Approximately 3   Approximately 5

10/11 Decimals
Partial answers

0.25   0.25
0.75   0.75

Reward stickers
When a page is completed and the answers checked, reward yourself with the right sticker.

12/13 Percentages and averages
Island map   Tourist numbers

2%   15%   12%
6%   89%   13%   15%
            20%   5%
            25%   10%

14/15 Kite flying game
Counters

14/15 Kite flying game
Calculation stickers

4 ÷ 2 = 2	16 ÷ 4 = 4	48 ÷ 6 = 8	99 ÷ 9 = 11
8 ÷ 2 = 4	20 ÷ 4 = 5	35 ÷ 7 = 5	22 ÷ 11 = 2
16 ÷ 2 = 8	28 ÷ 4 = 7	42 ÷ 7 = 6	33 ÷ 11 = 3
18 ÷ 2 = 9	32 ÷ 4 = 8	77 ÷ 7 = 11	44 ÷ 11 = 4
9 ÷ 3 = 3	40 ÷ 4 = 10	24 ÷ 8 = 3	55 ÷ 11 = 5
12 ÷ 3 = 4	25 ÷ 5 = 5	40 ÷ 8 = 5	66 ÷ 11 = 6
15 ÷ 3 = 5	30 ÷ 5 = 6	56 ÷ 8 = 7	24 ÷ 12 = 2
21 ÷ 3 = 7	45 ÷ 5 = 9	27 ÷ 9 = 3	48 ÷ 12 = 4
30 ÷ 3 = 10	18 ÷ 6 = 3	54 ÷ 9 = 6	60 ÷ 12 = 5
36 ÷ 3 = 12	36 ÷ 6 = 6	63 ÷ 9 = 7	72 ÷ 12 = 6

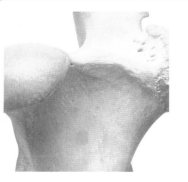
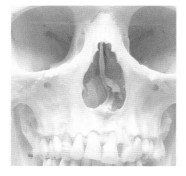

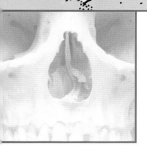

Approximation

Once you feel confident using long division, learn how to approximate. Approximation is knowing roughly what an answer to a calculation will be before doing it.

Long division practice
Study the key below and then practise long division by solving these problems. There will be a remainder in some of these calculations.

Bony facts
The human body has:
- 206 named bones
- 26 vertebrae, or back bones, in a spinal column
- 24 ribs in a rib cage
- 27 bones in each hand
- 26 bones in each foot

An archaeologist has found some bones. Most fit to make complete skeletons. Help her sort them out.

There are 336 ribs. How many complete skeletons could these fit into? ☐

The 864 hand bones go to make up ☐ pairs of hands.

886 is the foot bone total.
That is ☐ pairs of feet
☐ spare bones.

There are 361 vertebrae.
That is ☐ complete spinal columns
☐ spare bones.

Approximation
Knowing how to approximate helps you to check a long division calculation.

> 60 ÷ 19 is approximately 60 ÷ 20 = 3
> 60 ÷ 19 is approximately 3

Write down the approximate answers to these puzzles.

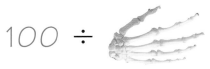

100 ÷ = ☐ Approximately

Number of hand bones

250 ÷ 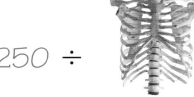 = ☐ Approximately

Number of ribs

Missing numbers
Find the stickers for these approximate answers.

62 ÷ 19 = ☆

39 ÷ 19 = ☆

149 ÷ 30 = ☆

What are the missing numbers in these approximations?

60 = 3010 ÷ ☐
Approximately

99 ÷ ☐ = 5
Approximately

9

Decimals

 Some calculations do not give a whole number answer. When you use a calculator, you may get a decimal in the answer. Solve these space puzzles to learn how decimals work.

Calculator decimals
Remember that the numbers after a decimal point are parts of a whole number, and the first shows tenths, the next hundreths, the next thousandths and so on.

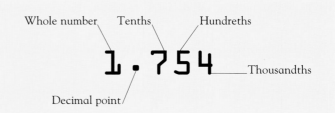

Calculator practice
Use a calculator to solve these calculations. Write in what appears on the calculator display.

50 ÷ 0.5 = ☐

500 ÷ 0.2 = ☐

426 ÷ 12 = ☐

785 ÷ 25 = ☐

Large decimals
Study the numbers below and see which is the biggest and which is the smallest. Fill in the boxes marking the biggest number with **1**, the middle number with **2**, and the smallest number with **3**.

50.11 ☐

50.9 ☐

50.01 ☐

Interplanetary travel
Some planets are light years away. This time cruncher machine reduces the journey time.

Light years

16.6 →

125 →

0.32 →

÷ 2

Fill in the boxes to show how long the journeys will take now.

Light years

☐ → ÷ 100 → 0.62

☐ → ÷ 100 → 0.05

☐ → ÷ 100 → 7.3

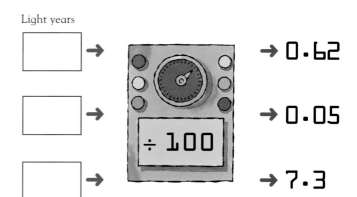

Fill in the boxes to show how many light years were put into the time cruncher machine.

Space buggy

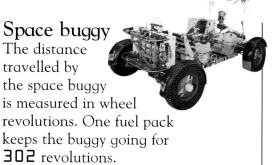

The distance travelled by the space buggy is measured in wheel revolutions. One fuel pack keeps the buggy going for 302 revolutions.

453 ÷ 302 = ☐
Revolutions travelled | Revolutions per pack | Number of fuel packs

6795 ÷ 302 = ☐
Revolutions travelled | Revolutions per pack | Number of fuel packs

How far will the buggy travel on 16 fuel packs? ☐

The buggy's performance changes with the terrain. Fill in these divisions to see how far it travels per fuel pack.

6000 revolutions on 6.25 fuel packs

☐ ÷ ☐ = ☐

240 revolutions on 0.75 fuel packs

☐ ÷ ☐ = ☐

Astronaut supplies

This astronaut's snack bag contains 8 snacks. What proportion of the snacks does he eat each time he has 1 snack?

1 ÷ 8 = ☐   Remember to write the answers as decimals.

If he eats 3 snacks what proportion of the full bag has he had? ☐

Planet samples

A lump of rock is brought back from space and divided equally amongst 400 laboratories. Fill in the boxes to show the rock parts in decimals.

1 part is   0.0025

5 parts are   ☐

10 parts are ☐

100 parts are ☐

40 parts are ☐

80 parts are ☐

Partial answers

Which of these calculations has an answer of 0.25? Sticker the correct answers.

23 ÷ 92 =

88 ÷ 22 =

18 ÷ 72 =

Which of these calculations have divisors of 0.75? Sticker the correct answers.

150 ÷ ☐ = 200

75 ÷ ☐ = 100

30 ÷ ☐ = 50

Percentages and averages

Sunny Island's tourist office produces many useful statistics for tourists visiting the island. Study these statistics and learn about percentages and averages.

Island map

This map comprises 100 squares. 6 of the map squares show sand-dunes. We can say 6 out of 100 squares are sand-dunes. Per cent means out of 100, so 6 per cent of the map is sand-dunes.

Study the map key below and count the number of squares on the map that show these symbols:

[ ] Squares   [ ] Squares

Now, write in the boxes below how many per cent of the map this is.

[ ] Per cent   [ ] Per cent

The mathematical sign for per cent is %. Write these numbers as percentages.

12 out of 100 _____

25 out of 100 _____

Map key

 Sand-dune   Forest   Airport

 Marshland   Town

Practise calculating percentages by doing these puzzles below and finding the correct sticker answers.

What percentage of the map is covered in marshland?

What percentage of the map is not covered in forest and marshland?

What percentage of the map contains airports?

What percentage of the map is covered in sand-dunes and towns?

Tourist numbers
These are the island's visitor numbers for each day in a week.

Monday	600
Tuesday	150
Wednesday	750
Thursday	450
Friday	300
Saturday	360
Sunday	390

 Use a calculator to find the percentage of tourists arriving each day.

This is how it is done:

| Number of tourists in 1 day | ÷ | Number of tourists in 1 week | × | 100 to give a percentage |

Find the correct sticker answer for each calculation.

Monday ☆	Tuesday ☆
Wednesday ☆	Thursday ☆
Friday ☆	Saturday ☆
Sunday ☆	

Average island weather
Here is a record of the number of sunny days the island had each month last year.

To find the average number of sunny days for each month, do the calculation shown below.

204 ÷ 12 = 17
Total number of sunny days | Total number of months | Average number of sunny days each month last year

What was the average number of sunny days a month from May to September? ☐

Average rainfall
The graph below shows last year's rainfall on Sunny Island.

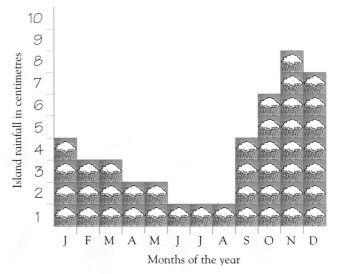

What was the average rainfall per month? ☐

Which 4 months are closest to the average?

_____

Of the total annual rainfall, what percentage falls in October, November, and December? ☐

13

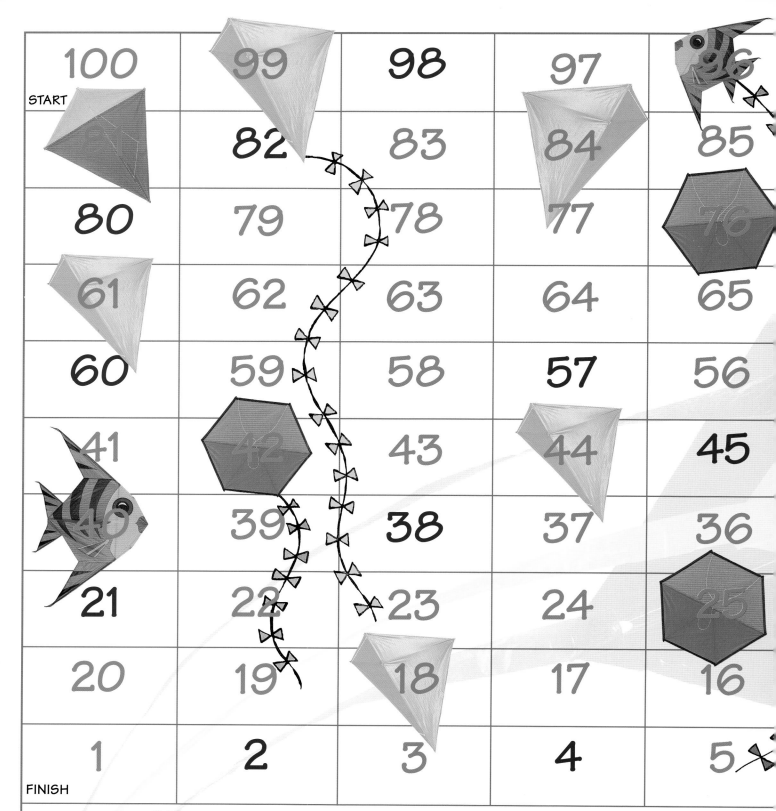

Kite flying game

You will need
- Two or more players
- Thin card
- Scissors
- 1 dice
- Counter and calculation stickers
- from the sticker sheet
- Scrap paper to work out the answers

Making the game pieces
Stick the counter and calculation stickers on to thin card and cut them out.

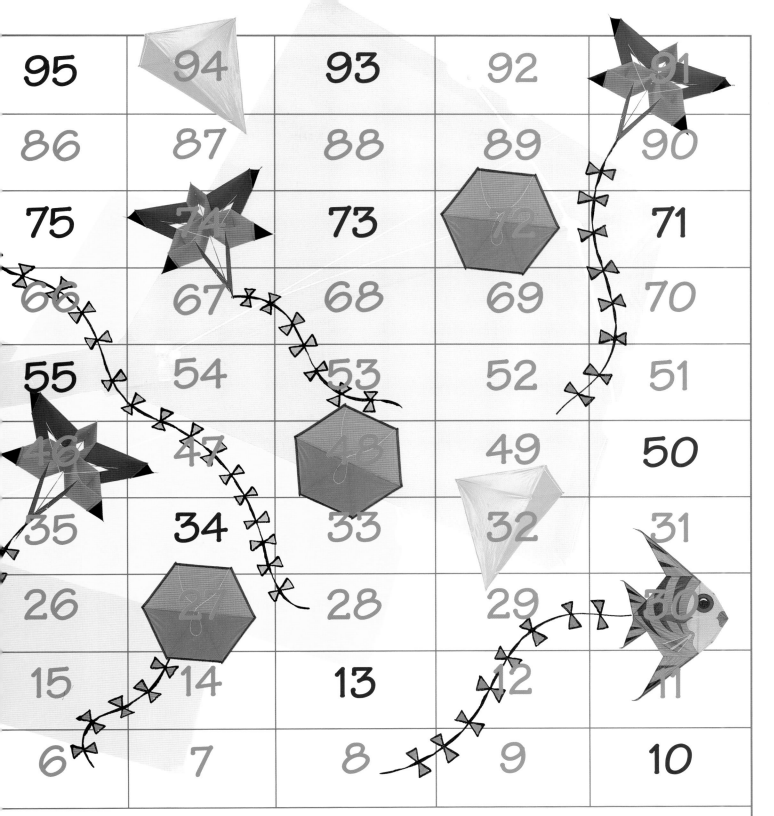

Playing the game
- Players each put a counter on the start square numbered 100 and the calculation cards face down in a pile.
- Take it in turns to throw the dice and move your counter along the board the number shown.
- If you land on a green number, your opponent reads out a calculation from the pile of calculation cards. If you answer the calculation correctly, have another go.
- If you land on the end of a kite tail, move your counter back up to where the kite is.
- The winner is the first person to get to the finish square numbered 1.

Answers

Pages 2/3 Dividing numbers
Dividing groups of fish
In 3 aquariums there will be **6** fish in each one:
$18 \div 3 = 6$
If half the fish are sold, **9** will be left:
$18 \div 2 = 9$
If the fish are shared out between 6 children, each child will receive **3** fish:
$18 \div 6 = 3$
If the fish are sold in pairs, there will be **9** pairs: $18 \div 2 = 9$
If 3 fish are sold, **5** fish will be left in each aquarium:
$18 - 3 = 15 \qquad 15 \div 3 = 5$

Leaping fish
$28 \div 7 = 4 \qquad 12 \div 3 = 4$
$30 \div 10 = 3 \qquad 48 \div 8 = 6$

Shellfish supper
10 shellfish each for **6** lion fish: $60 \div 10 = 6$
30 shellfish each for **2** lion fish: $60 \div 30 = 2$
6 shellfish each for **10** lion fish: $60 \div 6 = 10$
12 shellfish each for **5** lion fish: $60 \div 12 = 5$
20 shellfish each for **3** lion fish: $60 \div 20 = 3$

Division practice
$30 \div 5 = 6 \qquad 16 = 32 \div 2$
$35 \div 7 = 5 \qquad 12 = 24 \div 2$
$4 = 36 \div 9 \qquad 4 = 16 \div 4$

Pages 4/5 Division practice
Team players
66 players can go into **6** hockey teams:
$66 \div 11 = 6$
66 players can go into **11** volleyball teams:
$66 \div 6 = 11$
55 players can go into **11** basketball teams:
$55 \div 5 = 11$
55 players can go into **5** hockey teams:
$55 \div 11 = 5$

Tennis balls
$12 \div 6 = 2 \qquad 6 = 36 \div 6$
$42 \div 6 = 7 \qquad 3 = 18 \div 6$
$24 \div 6 = 4 \qquad 9 = 54 \div 6$

Match tickets
$12 \div 6 = 2 \qquad 1 = 12 \div 12$
$12 \div 3 = 4 \qquad 3 = 12 \div 4$
$12 \div 2 = 6$

Sports equipment
$24 \div 2 = 12 \qquad 24 \div 3 = 8$
$24 \div 6 = 4 \qquad 24 \div 4 = 6$
$24 \div 8 = 3 \qquad 24 \div 6 = 4$

Basketball scores
Baskets	By each star player	No. of stars
12	÷ 6 =	2
18	÷ 6 =	3
8	÷ 2 =	4

Pages 6/7 Remainders
Stereo puzzles
$17 \div 4 = 4$ remainder $1 \qquad 14 \div 3 = 3$ remainder 2
$23 \div 5 = 5$ remainder $3 \qquad 25 \div 6 = 6$ remainder 1
$9 \div 2 = 2$ remainder $1 \qquad 35 \div 8 = 8$ remainder 3

Cassette tapes
16 tapes divided between **5** stereos remainder **1**.
19 tapes divided between **6** stereos remainder **1**.

Transporting stock
202 boxes will fit into **2** trucks remainder **2**.
403 boxes will fit into **4** trucks remainder **3**.
222 stereos will fit into **2** trucks remainder **22**.

Factory calendar
There are **4** complete working weeks in the month of September.
2 working days remain in this month.
There are **4** complete working weeks in a 31-day month.
3 working days remain in a 31-day month.

Working week
6 days a week for 4 weeks: $6 \times 4 = 24$
$24 \div 5 = 4$ full weeks and **4** extra days.
5½ days a week for 4 weeks: $5½ \times 4 = 22$
$22 \div 5 = 4$ full weeks and **2** extra days.
5 days a week for 2 weeks and 5½ days a week for 2 weeks: $5 \times 2 = 10, \; 5½ \times 2 = 11$.
$10 + 11 = 21$
$21 \div 5 = 4$ full weeks and **1** extra day.

Page 8 Long division
Dig divisions
$559 \div 13 = 43$
$350 \div 25 = 14$
$442 \div 17 = 26$

Page 9 Approximation
Bony facts
336 ribs could fit into **14** complete skeletons.
864 hand bones go to make up **16** pairs of hands.
886 foot bones: **17** pairs of feet and **2** spare bones.
361 vertebrae: **13** complete spinal columns and **23** spare bones.

Approximation
$100 \div 27$ hand bones = approximately **4**
$250 \div 24$ ribs = approximately **10**

Missing numbers
$62 \div 19$ = approximately **3**
$39 \div 19$ = approximately **2**
$149 \div 30$ = approximately **5**
Approximately $60 = 3010 \div 50$
$99 \div 20$ = approximately 5

Pages 10/11 Decimals
Calculator practice
$50 \div 0.5 = 100 \qquad 426 \div 12 = 35.5$
$500 \div 0.2 = 2500 \qquad 785 \div 25 = 31.4$

Large decimals
50.11   2
50.9   1
50.01   3

Interplanetary travel
$16.6 \div 2 = 8.3 \qquad 62 \div 100 = 0.62$
$125 \div 2 = 62.5 \qquad 5 \div 100 = 0.05$
$0.32 \div 2 = 0.16 \qquad 730 \div 100 = 7.3$

Space buggy
$453 \div 302 = 1.5 \qquad 6795 \div 302 = 22.5$
On 16 fuel packs the buggy travels **4832** revolutions.
$6000 \div 6.25 = 960 \qquad 240 \div 0.75 = 320$

Astronaut supplies
$1 \div 8 = 0.125$
$3 \div 8 = 0.375$

Planet samples
5 parts are $\qquad 5 \div 400 = 0.0125$
10 parts are $\qquad 10 \div 400 = 0.025$
100 parts are $\qquad 100 \div 400 = 0.25$
40 parts are $\qquad 40 \div 400 = 0.1$
80 parts are $\qquad 80 \div 400 = 0.2$

Partial answers
$23 \div 92 = 0.25 \qquad 150 \div 0.75 = 200$
$18 \div 72 = 0.25 \qquad 75 \div 0.75 = 100$

Pages 12/13 Percentages and averages
Island map
5 forest squares = **5** per cent
9 town squares = **9** per cent
12 out of 100 = **12%**
25 out of 100 = **25%**
6% of the map is marshland.
89% of the map is not forest and marshland.
2% of the map contains airports.
15% of the map is sand-dunes and towns.

Tourist numbers
Monday: $600 \div 3000 \times 100 = 20\%$
Tuesday: $150 \div 3000 \times 100 = 5\%$
Wednesday: $750 \div 3000 \times 100 = 25\%$
Thursday: $450 \div 3000 \times 100 = 15\%$
Friday: $300 \div 3000 \times 100 = 10\%$
Saturday: $360 \div 3000 \times 100 = 12\%$
Sunday: $390 \div 3000 \times 100 = 13\%$

Average island weather
The average number of sunny days from May to September was **26**:
$130 \div 5 = 26$

Average rainfall
The average rainfall per month is **3.5 cm**:
$42 \div 12 = 3.5$ **cm**
The 4 months closest to the average are: **January, February, March**, and **September**.
50% of the year's rainfall fell in October, November, and December:
Total amount ÷ Total rainfall × 100 = **50%** of rainfall in Oct., Nov., and Dec. in 1 year
$21 \div 42 \times 100 = 50\%$